GROSS
SCIENCE EXPERIMENTS

by Q. L. Pearce

illustrated by Tim Harkins

LOWELL HOUSE JUVENILE

LOS ANGELES

CONTEMPORARY BOOKS

CHICAGO

To Cheryl, who can find humor in just about anything.
—Q. L. P.

Publisher: Jack Artenstein
Director of Publishing Services: Rena Copperman
Executive Managing Editor, Lowell House Juvenile: Brenda Pope-Ostrow
Editor in Chief, Lowell House Juvenile: Amy Downing
Director of Art Production: Bret Perry
Art Director: Lisa-Theresa Lenthall
Typesetter: Carolyn Wendt
Cover Art: Terry Kovalcik

Library of Congress Catalog Card Number: 97-14892

ISBN: 1-56565-594-X

Lowell House books can be purchased at special discounts when ordered in bulk for premiums and special sales. Contact Department TC at the following address:

Lowell House Juvenile
2020 Avenue of the Stars, Suite 300
Los Angeles, CA 90067

Manufactured in the United States of America
10 9 8 7 6 5 4 3 2

CONTENTS

GET GROSSED OUT ON SCIENCE!

What do you call someone who can bend bones in his or her bare hands, change a foul-smelling liquid from putrid purple to putrescent pink with a drop of acid, make glass groan, or rear colonies of creepy-crawly worms in a kitchen cupboard? A scientist, of course!

If these activities sound like your cup of goo, this collection of slippery, slimy, freaky, and fun experiments is for you. They are guaranteed to offend the senses while demonstrating cool science principles.

The experiments are grouped by the odors they secrete, the sickening sounds they make, or the yucky way they feel, taste, or look. Once you perform each test, read how it worked in the "Getting a Clue" section at the end of the experiment. Boldface words are defined in the glossary at the back of the book.

These experiments are worthy of an eerie hilltop laboratory, but most can be performed in a small amount of space with ingredients you may already have around the house. Also, an approximate time frame is given for each experiment, so you'll know how much time to allow.

But before you don that mad-scientist lab coat, read these *important* safety procedures!

- Before beginning *any* experiment, always read the instructions completely and get an adult's permission.

- If the experiment suggests asking an adult for help, do so! If you need to use heat or sharp objects in the test, be sure to ask a grown-up to give you a hand.

- *Never* taste anything unless the instructions specifically say it's okay. (Remember what turned Dr. Jekyll into Mr. Hyde?)

- Clean up everything when you are finished, and be sure to wash your hands before and after each experiment.

Ready to begin? If you have a strong stomach, you'll love doing these experiments. If you're squeamish, that's okay: Science isn't always pretty.

Here are some freakish displays of science at work that can be downright disgusting to watch!

BLOATED RAISIN BUGS

Time Required:
6 to 8 hours to complete

In just a matter of hours, you can turn a handful of innocent raisins into a herd of what look like bloated water bugs. Then you can really freak out your friends by eating one!

LET'S DO IT!

1 Fill the glass jar within 2 inches of the top rim with lukewarm tap water.

2 Drop the raisins into the jar, then place the jar where it can be left undisturbed. Let the experiment sit overnight or for at least 6 hours.

When the time is up, check your results. Instead of being tiny and shriveled, the raisins are now fat and bloated. Reach in and squeeze a couple. How did they get so heavy and squishy?

GETTING A CLUE

You are observing an example of **osmosis.** Water naturally moves from areas of greater **concentration** to areas of lesser concentration. In this case, the water travels through a **membrane,** the raisin skin. When you first put the raisins in the jar, lots of water was outside of each dried-up raisin and not much inside. The water moved through tiny openings in the raisin skin deep into the raisin itself, making it heavy, soft, and bloated.

ICKY EXTRA EXTRA

If you sit in a tub of water for a long time, your fingers probably will get wrinkly. Why doesn't your whole body undergo the same change? While you're in the tub, your skin will absorb some water, but much of your skin is partly protected from absorbing water by an oily substance called *sebum.* This substance is produced by tiny glands known as *sebaceous glands.* They occur all over the body, except in the skin of your fingertips and toes. As a result, water enters those areas by osmosis, and the skin there becomes wrinkly like a prune and puffy.

Try to figure out a way to prevent osmosis from taking place in this experiment. *Hint:* As sebum helps keep water from being absorbed through our skin, think of a substance that could do the same for the raisins. How about dipping the raisins in oil or painting them with clear nail polish before you begin? Try it and see!

JEEPERS, CHECK THOSE CREEPING PEEPERS

Time Required:
Approximately 1 week to complete

ADULT SUPERVISION NEEDED WHEN USING KNIFE

The small spud sits quietly in a corner. But what's happening? Are those long, spindly tentacles sprouting out of its eyes? It's the attack of the killer potato!

STUFF YOU'LL NEED

potato, any kind with eyes

knife

clean jar or glass with opening wide enough to fit potato

4 toothpicks

water

LET'S DO IT!

1 Check your potato for firmness. It should be hard with no dark, soft, rotten spots. With an adult's help, slice off the lower third of the potato. (You can dispose of it or save it for a smaller version of this experiment.) Stick four toothpicks securely in the larger piece, spacing them evenly around the potato about ½ inch above the bottom flat end.

2 Fill the jar nearly to the rim with cool tap water. Place the cut end of the potato in the water, with the toothpicks resting on the jar's opening.

3 Put the jar in a warm place that gets sunlight for at least an hour or two each day. Check every other day to be sure the water continues to cover the cut end of the potato.

4 In about a week the potato will sprout leaves and roots from little depressions known as *eyes*. If you want to keep your new plant, let it grow to about 6 inches tall, then plant about an inch of the root end of the potato in a pot of soil. Place it in a sunny location, keeping the soil moist, and your pet plant will flourish.

GETTING A CLUE

Vegetative propagation, demonstrated in this experiment, is the term for growing a new plant from a part of an adult plant rather than from seeds or **spores.**

Although a potato grows underground, it is not a root or a seed. It is actually a tuber, or food storage unit, for the potato plant.

FREAKY FACT

The world's largest potato was reportedly unearthed in 1795. It was an 18-pound, 4-ounce spud lurking in the soil of an English garden.

DisGUSTiNG **E**TAiLS

If you wanted to grow potatoes in your garden, you probably would not plant seeds, because potato plants grown from seeds can vary tremendously. Instead you would plant small pieces of potato called sets. *Each set must include at least one eye. The plants grown from pieces of potato are identical to the parent plant. It's almost like cloning in your own backyard!*

A tuber is the enlarged end of an underground stem. When conditions are right, the eyes on the potato can sprout leaves and roots, growing into a new plant. At first, the plant is nourished by the "meat" of the potato (the part we eat), which is mostly water, starch, and some **protein.** Once the new young plant is well established with a system of leaves and roots, it is nourished by the soil that it is planted in.

MERRY MAGGOT MENAGERIE

Time Required:
Phase 1 – 4 or 5 days to complete
Phase 2 – Approximately 2 weeks to complete

Who couldn't use a few more friends? Here's a way to raise your own ghastly brood, then watch as they change right before your very eyes. This experiment should be started on a nice spring or summer day when fruit flies are plentiful.

LET'S DO IT!

Phase 1:

1 Place the soon-to-be-rotten fruit in the open jar and put it in a safe, shady spot outside where it can sit undisturbed for at least 4 or 5 days.

2 The day after you start your experiment, check the jar. If you see tiny black fruit flies bouncing around on the fruit, cover the top with a piece of cheesecloth or panty hose and secure with a rubber band. If no fruit flies are spotted, continue checking the jar each day until they appear.

Once you've covered the jar, give the flies 2 days to enjoy their reeking meal and lay eggs. Then release them from the jar.

Phase 2:

 Re-cover the jar with the cheesecloth or panty hose and store it in a warm place where no people or animals will touch it.

 Over the next 2 weeks, check the fruit in the jar every day with a magnifying glass without removing the cloth covering. With luck, you'll see a colony of fruit flies form and observe their progression from eggs (tiny, grayish specks) to adults.

GETTING A CLUE

Leaving ripening fruit out is like sounding a dinner bell for adult fruit flies. Not only do they gorge themselves on the sweet pulp, but they lay eggs there, too, to give their young a meal the moment they hatch. Like many other insects, fruit flies go through four stages of development.

The egg is the first stage. The second stage is the larva, which looks something like a small worm. In the fly, the larva is generally called a *maggot*. The third, or pupa stage, is a resting stage during which the insect develops its adult features. Finally, during the fourth stage, an adult fly emerges. The entire process is called **metamorphosis**. Different kinds of insects go through the four stages of metamorphosis at different rates. For example, certain types lay eggs that hatch in hours, while others take days.

Disgusting Details

When police find a dead body, insects can serve as natural clocks in figuring out the time of death. The people who study this are called forensic entomologists. *They often can determine the time of death within minutes due to insect development on the body. When someone dies outside, within 10 minutes common green flies arrive to feed on flesh. They lay thousands of eggs, which hatch into maggots 12 hours later. Within 24 to 36 hours beetles arrive. Bug behavior also can help police determine whether the person was killed inside or out, at night or during the day, in warm or cold weather, and in sun or shade.*

ON PINS AND NEEDLES

Time Required:
Approximately 24 hours to complete

These crazy crystals really make a point, and the weird, crusty growths seem to appear out of nowhere. How do they form? Why are they shaped the way they are?

LET'S DO IT!

STUFF YOU'LL NEED

large jar lid

pencil

black construction paper

scissors

4 or 5 tablespoons Epsom salts (available at drugstores and many supermarkets)

1 cup lukewarm water

mixing spoon

1 Use the jar lid to draw a circle with a pencil on the construction paper. Cut out the circle and place it inside the jar lid. This will give the lid a dark background, making it easier to observe the crystals.

2 Add the Epsom salts to the cup of lukewarm water. Stir until the Epsom salts disappear.

3 Pour about ¼ inch of the salty water from the cup into the lid. Set the lid in a warm place with good air circulation, but not in direct sunlight. Leave it undisturbed for at least 24 hours. Long, thin, needlelike crystals will develop on the lid.

FREAKY FACT

For more than 300 years people have relied on a hot bath with Epsom salts to ease sore muscles. This common mineral is also known as *epsomite,* and it is often found as a scabby crust in mines and caves. The pure quality used in medicinal Epsom salts is named for a town called Epsom, in Surrey, England. There, evaporated mineral waters left behind huge deposits of the crystals, making the place a gigantic version of the experiment you just performed.

GETTING A CLUE

A crystal is a basic form of a solid substance in which the **molecules** are arranged in a regular, geometric pattern. When solid substances are pure (made up of a single kind of molecule), they usually have a definite crystal form that has flat surfaces that always meet at the same angle.

The Epsom salts used in this experiment were crystals before they were ground into a grainy powder. When the powdered Epsom salts dissolve in water, they form a solution in which the salt molecules are randomly scattered. As the water **evaporates**, it leaves the salt behind. The salt molecules return to their natural orderly pattern, forming thin crystals. That is the way they looked before being crushed into grainy Epsom salts. The formation of crystals is called *crystallization.*

DISGUSTING DETAILS

There are many salt lakes in the world—bodies of water so high in salt content that as the lake waters evaporate, salt crystals are left behind on the shoreline. One very awesome salt lake is Mono Lake. Located in California, it is three times as salty as the Atlantic Ocean. It is also loaded with sodium carbonate, a close cousin to baking soda. This unique combination supports a growth of blue-green algae that acts as food for an outrageously gross population of brine flies. At the end of the summer some 4,000 flies crowd each square foot of salt-encrusted shoreline! In the late 1800s that meant good eating for members of the native Paiute tribe. They gathered and ate the larva and pupa forms of the fly, which are yellowish and oily but very nutritious.

MUGGY MUDDY WATERS

Time Required:
Approximately 24 hours to complete

Create a murky swamp, then clean it up without even breaking a sweat. (Just wait until you see the muddy mess this experiment leaves behind!)

STUFF YOU'LL NEED

2 cereal bowls
water
2 tablespoons of dirt
mixing spoon
brick (or shoe box)
white handkerchief

LET'S DO IT!

1 Fill one cereal bowl halfway with water, then add 2 tablespoons of dirt and stir well.

2 Put the brick (or upside-down shoe box) on a flat surface and set the bowl of muddy water on top of it. Place the empty second bowl on the flat surface next to the brick, so it rests several inches below the muddy water.

3 Roll the handkerchief into a tube, tight enough to stay rolled up. Place one end of the handkerchief in the muddy water and the other end in the empty bowl, not quite touching the bottom.

4 Wait 24 hours, then check the experiment. How did the water get into the once-empty bowl? Where is the dirt?

GETTING A CLUE

You are looking at an example of **capillary action.** Say what? It's like this: The cloth is made up of tiny fibers with air spaces in between. The water slowly creeps up into the spaces, a little higher on the sides than in the middle, giving the water level a cup shape. The water molecules attract each other, soon drawing the center of the water surface up the cloth, causing the entire waterline to become level. When that happens, the water on the sides creeps up again and so on, until the water makes it over the edge of the bowl. Gravity takes over and the water slowly seeps down the hanky, finally dripping into the lower bowl. The dirt is left behind because it's too heavy to hitch a ride up the hanky.

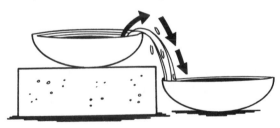

Important! No matter how clean the water in the lower bowl looks, don't drink it. The dirt may have been too heavy to make the trip, but lightweight **bacteria** might have hitched a ride and could make you sick if you drink the water.

FREAKY FACT

Water rises in plant stems and tree trunks by capillary action. For some trees, the journey of the water to the top is a pretty impressive trip. For example, the tallest tree in the United States stands in Redwood National Park, Orick, California, and measures more than 365 feet tall.

17

MORBID BLEEDING BLOBS

Time Required:
Less than 1 hour to complete

ADULT SUPERVISION NEEDED FOR ENTIRE EXPERIMENT

These drippy spheres are guaranteed to make you gasp. What makes them bob? How do they bleed?

LET'S DO IT!

1 Fill the jar halfway with rubbing alcohol. Add 5 teaspoons of water. Stir the mixture with a spoon, then set the jar aside and wait for the liquid to stop moving.

2 Meanwhile, place 4 teaspoons of vegetable oil into the paper cup. Add 2 drops of red food coloring, then use the eyedropper to quickly stir the food coloring into the oil. The food coloring stays separated from the oil in tiny balls. Keep mixing and the oil, as it turns blood-red, will look like it blends completely with the food coloring.

3 Fill the eyedropper with the food coloring and oil mixture. Carefully lower the eyedropper into the jar of alcohol and water solution and place the tip just below the surface. Squeeze the end of the dropper to release a blob of colored oil. Take the eyedropper out of the jar and squeeze out the extra oil onto a paper towel. Refill the eyedropper with more colored oil and repeat the process until you have three or four blobs in the jar.

4 Watch the result. Your goal is to have the blobs floating in the middle of the jar. But your blood-red blobs will probably do one of two things:

- They will sink. (If they settle at the bottom of the jar, you can add more water, a teaspoon at a time, to make the blobs rise.)

- They will float. (If they rise to the top, you can add more alcohol, a teaspoon at a time, to make the blobs sink.)

Once in a great while, one kid in about 8 million will mix the alcohol and water perfectly on the first try, making the red blobs settle in the middle of the jar. For the rest of you, as you patiently add water or alcohol, teaspoon by teaspoon, watch this process from the side of the jar. *Don't* place your face near the top of the jar—it puts you in contact with harmful fumes.

When the blobs are in position, put the lid back on the jar. Peer carefully at the blobs. How do they stay where they are? Why do they look like they're bleeding?

Important! Once you've finished admiring your bleeding blobs, immediately pour the liquid down the sink, then run tap water for a minute or two. Do not leave it standing around where anyone, especially small children, can get to it. Rubbing alcohol is highly **toxic** if swallowed.

GETTING A CLUE

There are a couple of things going on in this experiment. The first is a demonstration of **density**, the ratio of weight to **volume** in an object or substance. The oil-based blobs bob in the middle of the jar because the alcohol and water mixture is lighter than water yet heavier than alcohol. In a perfectly mixed **solution**, the blobs are somewhere in between, density-wise—they neither sink to the bottom nor float to the top of the jar.

This experiment is also an example of **immiscible liquids**, or liquids that do not mix together. The oil stays in a round blob because it won't mix with water or alcohol. The blob "bleeds" because the food coloring mixes better with water and alcohol than it does with oil, making food coloring and oil immiscible liquids, too.

DISGUSTING DETAILS

There is an old saying that oil and water do not mix. In science lingo, they are immiscible liquids, and there have been many tragic examples of this in the earth's oceans. Anyone who has seen or smelled an oil slick knows just how disgusting it can be. Oil from shipping accidents floats on top of ocean water and eventually makes its way toward the shore, where it leaves an ugly black sludge. To clean up an oil slick, some countries use a chemical spray that makes the oil sink deeper in the water, keeping it away from shore. Others clean up an oil slick by spraying detergent, which breaks up the oil into smaller particles that are then more readily broken up by ocean **microorganisms**.

ICKY EXTRA EXTRA

Why not try adding some detergent to your experiment to see what happens? (A few drops of dish detergent will do.)

DiD YOU HEAR THAT?

Clean out that earwax buildup! Otherwise, you won't believe your ears as you tap, rattle, and scrunch your way through these disgusting experiments.

MOANING MARVIN

Time Required:
Less than 1 hour to complete

ADULT SUPERVISION NEEDED WHEN DRILLING OR HAMMERING HOLE IN WOOD

When Marvin gets going, the whine of his wooden head can send chills up and down your spine.

LET'S DO IT!

STUFF YOU'LL NEED

rectangle of pine or other light wood, approximately 8 inches long, 2 inches wide, and ¼ inch thick

3-foot length of ⅛-inch-thick string

drill or hammer and large nail

colored markers

1 Ask an adult to help you make a hole in one end of the block of wood just large enough to pass the string through. Using a drill is the easiest way, but it also can be done with a hammer and a large nail.

2 Draw a scary face on the wood with the markers.

3 Pass the string through the hole and tie the end securely with double knots. Have an adult check your work—you don't want to launch Marvin into orbit in the middle of the experiment!

4 To perform the experiment you must move outside, at least 10 feet away from buildings, trees, and people. Grip the loose end of the string and whirl the wood in a large circle above your head. Check out the spooky moaning sound Marvin makes!

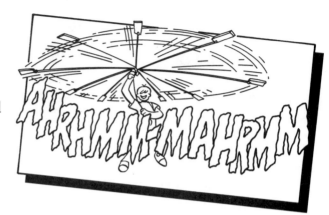

GETTING A CLUE

There are a couple of things at work here. The moaning **sound** is created because the wood and cord vibrate as they move through the air, causing the surrounding air to vibrate. These air vibrations travel in waves and eventually strike a membrane in your ear called your *eardrum.* The vibrations are then passed on to nerve endings in your inner ear and you hear the vibrations as sound. Because Marvin is moving in a circle, the vibrations repeat at intervals that make a sound like a moan.

The force that keeps Marvin going in a circle instead of flying off into space is called **centripetal force**. Objects tend to want to keep moving in a straight line, unless a force makes them change direction. Your arm and the string are applying that force to Marvin.

ICKY EXTRA EXTRA

If you slow down or speed up, you can change the **pitch** of the moan. Pitch is a measure of the **frequency** of vibrations. Try spinning Marvin at different speeds to see how the sound changes.

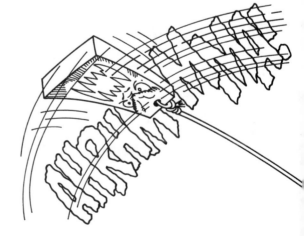

GROANING GLASS

Time Required:
Less than 1 hour to complete

These grating sound effects can set your teeth on edge. Try it with glasses of different sizes, and you can start your own creepy chorus.

LET'S DO IT!

1 Wash the glass and your hands in warm, soapy water to get rid of any grease or oil. Dry thoroughly.

2 Holding it by the stem, place the squeaky-clean glass on a flat surface. Pour a little vinegar into the bowl (you won't need more than ¼ cup) and dip the index finger of your free hand in the vinegar.

3 Holding the glass at its base, rub your wet finger around the rim of the glass until you hear a high-pitched whine.

GETTING A CLUE

Washing everything well and using vinegar rids the glass and your finger of any oil or **lubricant**. As you rub, you create **friction** between the rim of the glass and your finger. Friction is the resistance to motion between two surfaces moving across each other. This resistance causes the glass and surrounding air to vibrate, which you hear as a whining sound. The pitch of the sound is determined by the frequency or number of vibrations per second. The higher the frequency, the higher the pitch.

ICKY EXTRA EXTRA

The friction between your finger and the glass results in the sound you hear in this experiment. Another product of friction is **heat**. To feel heat from friction, try rubbing your dry hands together very fast. Do they warm up?

THE CRUSHER

Time Required:
Less than 1 hour to complete

What could be better than bending steel with your bare hands? How about destroying a bottle with a sickening crunch from 10 feet away?

STUFF YOU'LL NEED

approximately ½ cup warm tap water

1-liter plastic soda bottle with cap

10 ice cubes

plastic bag

small hammer or mallet

LET'S DO IT!

1 Pour the warm water in the bottle, then immediately screw on the cap and set aside.

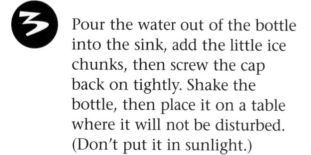

2 Place the ice cubes in the plastic bag and crush them into chilling chunks with the hammer or mallet.

3 Pour the water out of the bottle into the sink, add the little ice chunks, then screw the cap back on tightly. Shake the bottle, then place it on a table where it will not be disturbed. (Don't put it in sunlight.)

Now sit back and watch! It might take as little as 10 minutes before you hear the first crunch; then slowly, over the next several minutes, the bottle will cave in on itself, popping and groaning all the while!

GETTING A CLUE

Air pressure is the pressure that air in the atmosphere exerts on everything it touches. It is created by the weight of air above the earth's surface.

Empty plastic bottles don't usually collapse when they are just sitting around because the pressure inside and outside the bottle is equal. In other words, the air inside the bottle pushes outward as hard as the outside air pushes in. In this experiment you first warm the air in the bottle to cause it to expand, then you chill it rapidly. As the warm air in the bottle cools, it **contracts**, or shrinks. It takes up less space, so the air pressure on the inside of the bottle is lower than the outside air pressure. The higher pressure on the outside causes the bottle to cave in on itself.

DISGUSTING DETAILS

*At sea level and at 32°F, the earth's atmosphere presses down at about 15 pounds per square inch. At this very moment, your body is experiencing this pressure, but you don't collapse inward because the air and **fluids** inside of you push outward (or expand) until the forces pushing out and the forces pushing in are equal.*

That's okay if you are on this planet, but it could pose a problem if you were in outer space. On a space walk an astronaut has to wear a pressurized suit to keep from blowing up like a balloon. Without a pressurized suit, not only would the space traveler have nothing to breathe, but the air and fluids in his or her body would keep expanding until the space traveler's eyeballs popped out and his or her skin and other organs ruptured.

SNEAKY SNAKES

Time Required:
Less than 1 hour to complete

What could be scarier than hearing the rattle of a rattlesnake? Holding the source of that disturbing sound in your hands!

LET'S DO IT!

STUFF YOU'LL NEED

needle-nose pliers

piece of strong wire about 5 inches long (coat-hanger wire works well)

two 3-inch-long rubber bands

metal washer, approximately 1 inch in diameter

paper coin-envelope

pencil

an unsuspecting friend

1 Using the pliers, curl the two ends of the wire as shown. Make a small bend in the center of the wire, forming an open-ended triangle.

2 Thread one rubber band halfway through the hole in the washer, then hook both ends of the band onto one curved end of the wire.

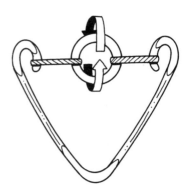

Repeat the process on the other wire end with the second rubber band, threading it through the same washer.

 Take the coin envelope and write *BEWARE: Snake!* in large letters. Now turn the washer about twenty times as you would turn the winder on a windup toy. Slip the entire contraption into the coin envelope, being careful not to let the bands unwind. Keep the washer flat and slip the flap of the envelope inside.

Hand your prepared envelope to a friend and say, "Open this envelope if you dare!" Your friend will have to reach in to pull out the flap. That will release the washer and it will spin, making a disturbing rattlesnake sound in the paper envelope.

GETTING A CLUE

This experiment is a demonstration of potential and kinetic **energy.** Energy is the ability to do work. **Potential energy** is the possible energy something has because of its position. When the position changes, the object can accomplish work. For example, a ball at the top of a staircase has potential energy. Give it a nudge, and as the ball tumbles down the stairs, its potential energy changes to **kinetic energy.** Kinetic energy is the energy of motion, or energy that is at work. In Sneaky Snakes, the washer and the tightly wound rubber bands have potential energy until the washer is released, then the spinning washer and unwinding rubber bands have kinetic energy.

FREAKY FACT

Energy can be stored in two main ways. One is the potential energy a body has because of its position. Another way energy can be stored is as **chemical energy,** or the energy stored in an **atom** or molecule, which can be released through a **chemical reaction.** Examples of stored chemical energy are gas in a car or food in your body. When you run around, the chemical energy in the food you have eaten is released. About 25 percent of it is changed to kinetic energy. Most of it is converted to heat so the body has to find ways to cool itself down. Which brings us to a gross side-effect of using up all this energy—sweat. In fact, if you lived in a hot climate and played hard all day, you could sweat out as much as 12 quarts of fluid in a day.

ALL IN BAD TASTE

These experiments will terrorize the taste buds and may make you lose your appetite.

A HAIRY SPONGE

Time Required:
Approximately 1 week
to complete

**This unusual garden
grows so fast it will
stand your hair on end.
But can it really survive
without soil?**

LET'S DO IT!

1 Place a handful of seeds in a
small bowl and cover them with water. Let the seeds soak for about
an hour, then pour them into a strainer to separate them from the water.

2 Soak the sponge for a few moments in water, wring it out, and place it
in the saucer.

3 Sprinkle a handful of the wet seeds from the strainer on the sponge and
poke them down into the openings.

4 Place the saucer in a warm area, but
not in direct sunlight. Using the spray
bottle, keep the sponge moist over
the next 2 or 3 days. The seeds will
soon sprout, and tiny hairlike plants
will shoot up. When your garden
needs a trim, just munch away. It
should stay fresh for a week or two.

GETTING A CLUE

Most plants make their own food with sunlight, air, and water in a process called **photosynthesis.** As they grow, they get other nutrients they need from the soil. When seed-bearing plants first begin to sprout, they are unable to use photosynthesis to produce food because their leaves are too immature to do so. Seeds, however, have a limited food supply inside them that helps get the new plants started. Once this food supply is used up, the seeds fall away from the new plants and decay. In this experiment the tiny plants in the sponge use the stored food in the seeds to get started. If you want to keep the plants going longer than a week or so, add a few drops of liquid fertilizer to the sponge or transplant them to your garden.

ICKY EXTRA EXTRA

Do you want to observe a tiny new plant while it is still inside the seed? If you soak a lima bean in water for 24 hours, you can carefully split it open and observe the baby plant, called an **embryo,** within.

FREAKY FACT

Seeds come in all sizes. The world's smallest seeds are those of certain orchids. They are so tiny that it takes more than 28 million of them to weigh an ounce. The heaviest seed is that of the coco-de-mer, found on the Seychelles Islands in the Indian Ocean. A single seed of this plant may weigh 44 pounds. Imagine trying to plant a row of those in your garden!

FOUL FRUIT

Time Required:
2 hours to complete

ADULT SUPERVISION NEEDED WHEN USING KNIFE

Imagine this: You have your taste buds all ready for a yummy apple. You open wide, take a big bite, and . . . disgusting! You get a mouthful of brown, mealy pulp. Here, you can speed up the process of fresh, crisp fruit turning to mush.

STUFF YOU'LL NEED

knife
ripe apple or banana
2 small bowls
lemon juice
mixing spoon

LET'S DO IT!

1 With an adult's help, peel the fruit and cut it into small pieces. Put half in one bowl and set it aside.

2 Put the other half of the fruit in the second bowl, add the lemon juice, and toss with a spoon until the fruit is coated.

3 After about 30 minutes, check the fruit in each bowl. The bare fruit should be starting to turn brown. The fruit with the lemon juice should look much fresher.

4 Leave the fruit undisturbed for 2 more hours, then check again. By this time the bare fruit should be quite brown. If you want to up the grossness quotient of this experiment, let the untreated fruit remain out in the bowl for several days and observe the changes it goes through.

GETTING A CLUE

Chemicals in the fruit, called **aldehydes**, combine with oxygen in the air causing a **chemical reaction.** A chemical reaction is a change that takes place when two or more substances interact. Here, the aldehydes in the fruit mix with oxygen, turning the bare fruit sickly brown. After several hours it begins to get mushy. The fruit coated with the lemon juice will also turn brown, but much more slowly, because lemon juice is a weak **acid** that delays the process. Such weak acids are called **antioxidants.** Antioxidants are used as preservatives in many foods, paints, and fuels.

ICKY EXTRA EXTRA

You can speed up the browning process dramatically by placing the fruit in a blender and adding about 4 tablespoons of hydrogen peroxide. Blending exposes more of the aldehydes to the air and the peroxide releases oxygen. Once you have watched the show, carefully dispose of the mess. It is not safe to eat! It could make you very sick and it tastes truly awful!

DEMENTED DESSERT

Time Required:
4 hours to complete

ADULT SUPERVISION NEEDED WHEN USING STOVE

In an old movie called *The Blob,* a creature that looked a lot like an oversized lump of gelatin gobbled up a bellyful of townsfolk. Usually it's the other way around—flavored gelatin is a great dessert that kids can really sink their teeth into. It also makes a gooey, oozy example of a rather unusual substance. Try this demonstration, then feel free to eat the evidence.

STUFF YOU'LL NEED

1 cup apple juice

saucepan

small glass bowl (Pyrex® is recommended)

1 packet unflavored gelatin

mixing spoon

butter knife

microwave-safe plate

LET'S DO IT!

1 Pour the cup of apple juice into the pan and warm it over medium heat until it comes to a boil. Ask an adult to remove the juice from the heat and pour it into a glass bowl.

2 Sprinkle in the unflavored gelatin. Stir until the powder dissolves and refrigerate for several hours.

3 Once the gelatin has set completely, remove it from the bowl in one piece. To do this, place the bowl in warm water for 2 minutes. Then run a butter knife around the edge of the gelatin. Slide the gelatin carefully onto a microwave-safe plate.

Now place the wiggly gelatin in a microwave on medium power for about 30 seconds, then check it. (If you don't have a microwave, you can use an oven heated on its lowest setting. You'll need to substitute an oven-safe plate for a microwave-safe plate.) You should see that the

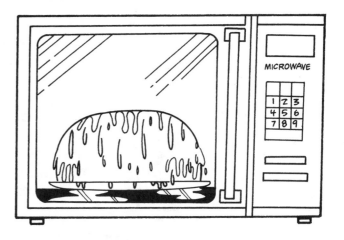

juice has been released and the slippery ooze is seeping across the plate. If your microwave is the type that has a clear door, you can even watch as the gelatin begins to "sweat."

GETTING A CLUE

Gelatin dessert is a substance known as a *colloid*. A colloid is made up of very tiny particles, called the *solute,* in a liquid, called the *solvent.* If the solute is solid and the solvent is liquid, then the colloid is known as a *sol* (such as milk and glue). If the solute and the solvent are both liquid, the colloid is called an *emulsion* (such as mayonnaise).

In this experiment, the colloid is called a *gel.* The solid part is the powdered gelatin—a protein that forms a kind of mesh. When mixed with hot water (or in this case, hot apple juice), gelatin granules absorb the liquid and swell, then melt. As the solution cools and is chilled, the liquid is trapped within the mesh network. The network a colloid forms is broken down by heat. When you put your gelatin in the microwave, the fluid was released because the network trapping it was no longer in place.

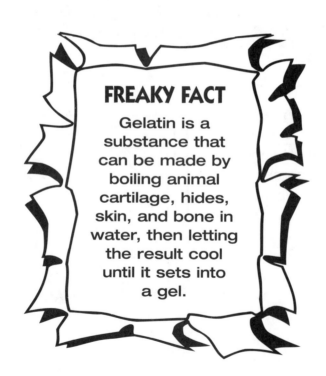

FREAKY FACT

Gelatin is a substance that can be made by boiling animal cartilage, hides, skin, and bone in water, then letting the result cool until it sets into a gel.

ALONG CAME A SPIDER

Time Required:
½ hour to complete

ADULT SUPERVISION NEEDED WHEN USING STOVE

Little Miss Muffet wasn't crazy about arachnids, but she loved curds and whey. Here's how to whip up a batch for yourself while demonstrating a chemical reaction called *protein coagulation.*

STUFF YOU'LL NEED

1 quart skim milk (1%, 2%, and whole milk also work)

saucepan

mixing spoon

2 tablespoons lemon juice

colander or sieve

cheesecloth

LET'S DO IT!

1 Pour the milk into the saucepan. With an adult's help, stir constantly over medium heat, bringing it to a slow boil. Keep a close watch on it because milk can boil over quite suddenly.

2 Have your adult helper remove the pan from the heat and stir in the lemon juice.

3 Return the pan to the heat and stir the milky mixture until you see lumps. Turn off the heat and let the milk cool to room temperature.

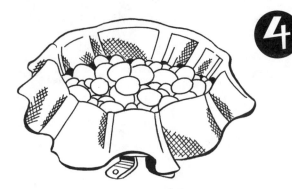

4 Line the colander or sieve with cheese-cloth and put it in the sink. Pour the milk through the colander. The pale fluid running down the drain is called *whey*. The lumps left behind in the colander are the curds, a simplified version of cottage cheese! It is certainly safe to eat, though you may not find it very tasty.

GETTING A CLUE

When a water-**soluble** protein, such as that in milk, is heated to a certain temperature, or acid is added, it becomes **denatured**. (In this experiment both heat and acid are used.) "Denatured" means a change takes place and the protein will no longer remain dissolved in a fluid. It forms clumps, or curds. The acid in the lemon juice causes such a chemical reaction when it comes in contact with protein in the milk. The result is curds and whey. The clumping reaction demonstrates protein coagulation. This is the process used to make cheese, a valuable source of protein in the human diet.

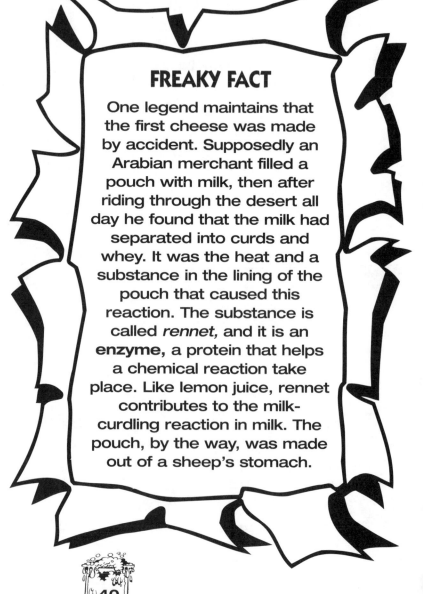

FREAKY FACT

One legend maintains that the first cheese was made by accident. Supposedly an Arabian merchant filled a pouch with milk, then after riding through the desert all day he found that the milk had separated into curds and whey. It was the heat and a substance in the lining of the pouch that caused this reaction. The substance is called *rennet,* and it is an **enzyme,** a protein that helps a chemical reaction take place. Like lemon juice, rennet contributes to the milk-curdling reaction in milk. The pouch, by the way, was made out of a sheep's stomach.

iCKY, STiCKY, SLiMy FUN

It's slime time.
The following experiments
put plenty of ooze, goop,
glop, and other cool
sensations right at
your fingertips.

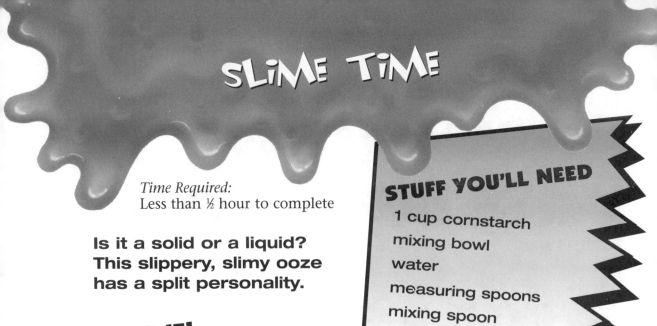

SLIME TIME

Time Required:
Less than ½ hour to complete

**Is it a solid or a liquid?
This slippery, slimy ooze
has a split personality.**

STUFF YOU'LL NEED

1 cup cornstarch
mixing bowl
water
measuring spoons
mixing spoon

LET'S DO IT!

1 Put the cornstarch in the bowl. Add water by the teaspoonful and stir. Keep adding teaspoons of water until the mixture is fluid but very hard to stir. It should have the consistency of thick mud.

2 Reach in and pick up a handful of the gooey fluid. Don't be squeamish—you have to move fast for this to work. Quickly roll the glop between your hands. It will become a firm ball.

3 Hold your hands over the bowl and stop rolling. The ball becomes fluid again and slippery slime drips through your fingers.

GETTING A CLUE

Your slime is an example of a non-Newtonian fluid. Fluids have a property called **viscosity**, or resistance to flow. The British scientist Sir Isaac Newton (1642–1727) is best known for his theories about light, gravity, and motion. He said that the viscosity of a fluid, such as water, can be changed only by raising or lowering its temperature. Sometimes in science you'll find an exception to a rule, and non-Newtonian fluids are the exception to Sir Isaac's rule on viscosity. Non-Newtonian fluids can be changed by temperature or, as shown in this experiment, by applying force. Putting pressure on the slippery slime by rubbing it between your hands makes it resistant to flowing.

FREAKY FACT

Believe it or not, glass is not a solid but rather a very viscous liquid. Adding heat will cause glass to flow. Non-Newtonian fluids, like glass, may also change over time. If you are lucky enough to live near a very old building that has the original glass in its windows, check closely. You may find that the panes are a tiny bit thicker at the bottom. That is because the glass has flowed ever so slightly downward over many years.

HERE COMES THE SLUDGE

Time Required:
Less than ½ hour to complete

Whip up some strange behavior in a bowl. What does it do? Find out . . . if you have the nerve.

LET'S DO IT!

1 With a spoon, mix the contents of the bottle of glue and ½ cup distilled water in one bowl. Add 5 drops of food coloring and stir well.

STUFF YOU'LL NEED

mixing spoon

4-ounce bottle of white glue

measuring cups

1 pint distilled water

2 medium-sized glass bowls

food coloring, any color, or mix up a particularly gross yellow-green

1 teaspoon borax powder

2 Pour 1 cup of distilled water and the borax powder in the second bowl. Stir well.

3 Pour the glue mixture into the borax mixture and keep stirring until you have a thick glob. Lift it out of the bowl and knead the mixture until it feels like dough. The behavior of the sludge depends on what you do to it, so get creative!
Rip the sludge in two. Find out what happens when you pull it slowly.
Will it bounce?

GETTING A CLUE

This is another slippery, slimy example of a non-Newtonian fluid. It is a concentrated **suspension.** That means it's a liquid with lots of tiny, solid particles suspended in it. The solid particles do not dissolve in the liquid. In fact, they can be removed by passing the liquid through a filter. You can try this yourself using a pint-sized glass jar and a number four coffee filter. Simply place the filter in the top of the jar and slowly pour the liquid into it. The solid material left behind in the filter is called a *residue*.

THEM BONES!

Time Required:
3 to 5 days to complete

Your friends and family will be amazed when they see you tying a bone in knots.

LET'S DO IT!

1 After a nice chicken dinner, save the bone you want to use for your experiment. With the scrub brush, carefully clean all the meat and gristle from the bone and allow it to dry for a few hours.

2 Fill the jar with vinegar, drop the bone in, and put on the lid. Leave the jar undisturbed for 3 days.

3 Remove the bone from the vinegar and rinse it with water. It should be soft and rubbery enough to tie in a knot. If it isn't, prepare a fresh jar of vinegar and leave the bone in it for another day or two.

GETTING A CLUE

Two main elements make up bone. Collagen is a protein that makes bones strong and resilient. Without it, they become very brittle. Apatite is a mineral that makes bones hard. The vinegar is a weak acid that dissolves the apatite in the bone but leaves behind the rubbery collagen.

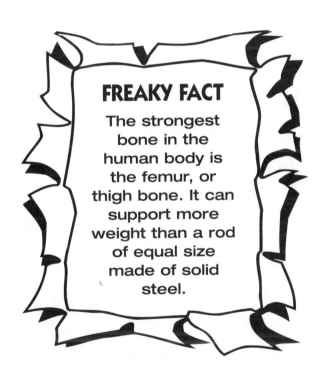

FREAKY FACT

The strongest bone in the human body is the femur, or thigh bone. It can support more weight than a rod of equal size made of solid steel.

NAKED EGG

Time Required:
3 days to complete

Have you ever peeled an egg . . . a raw egg, that is? There's a delightfully slimy surprise inside.

LET'S DO IT!

1 Put the egg in the jar, being very careful not to crack the shell.

2 Fill the jar with enough vinegar to completely cover the egg. Put on the lid and leave the jar undisturbed for about 3 days.

If you check once in a while, you'll see lots of tiny bubbles on the eggshell. When few bubbles are left and the shell is dissolved, very gently pour the vinegar out over the sink so you can catch the egg. Hold it very carefully and observe. It feels soft and slippery. Don't squeeze it or the egg will break.

Hold it up to a light. You can easily see the yolk inside. If you want to save the egg for a while, refill the jar with water, slip the egg in, and replace the lid. It will keep well in the refrigerator for a few days.

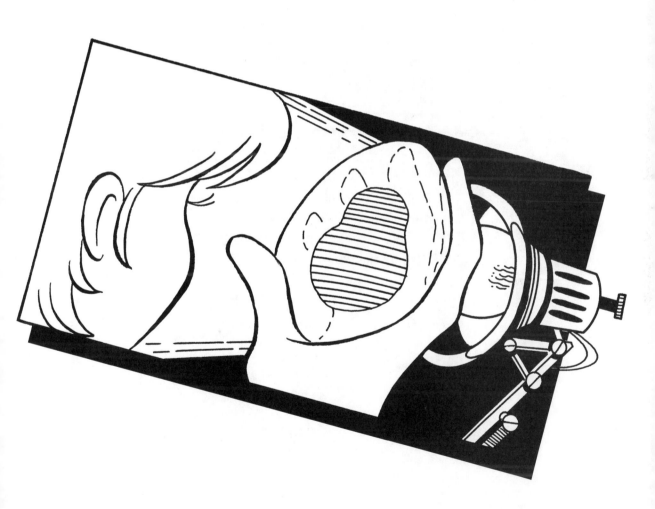

GETTING A CLUE

The eggshell is made up of a material called *calcium carbonate*. The vinegar reacts with the shell, dissolving it and producing tiny bubbles of carbon dioxide gas. Just inside the egg is a thin membrane that does not react to the vinegar. The membrane keeps the egg intact. Within is the yolk, which is meant to be the food supply for a developing embryo. This is surrounded by the egg white, or albumen, which helps cushion the embryo.

ICKY EXTRA EXTRA

If you store your shell-less egg in water, it will swell because of osmosis. (See page 6, Bloated Raisin Bugs, for an osmosis experiment.) Water moves through the egg membrane, but the molecules that make up the goopy material inside the egg are too big to go the other way. That's called *semipermeability*.

To see the opposite reaction, place the shell-less, raw egg in a jar of sticky corn syrup. There is more water in the egg than in the corn syrup, so the water molecules will leave the egg. The egg will shrivel into a gooey, yucky mess . . . er, mass.

MAKING A BIG STINK

These experiments will cause you to hold your nose . . . or may leave you gasping for fresh air!

A FUNGUS AMONGUS

Time Required:
1 week to complete

Here's an idea for a particularly disgusting gag lunch—a mold sandwich. No joke . . . the musty odor can really make you gag!

LET'S DO IT!

1 To dampen the slice of bread, pour the teaspoon of water on a flat plate and spread it around, getting as much of the plate wet as possible. Then lay the bread in it. If needed, sprinkle a little more water on the bread with a spoon.

2 Remove the moist bread from the plate, slip it into the sandwich bag, and fasten the bag shut.

3 Put the bag in a warm, dark place for one week—a closet or cupboard will work. Label the bag with a warning such as *experiment in progress* to let your family know that it is an experiment and not lunch. Whatever you do, don't forget about it!

4 Check once a day, every day, to see how the fungus is developing. If you wish, you can draw sketches or make notes about what you see so you have a record of the progress of the experiment. This will make it easier to see how quickly the fungus grows from day to day. At the end of the week you should find that the bread is covered with a layer of mold.

GETTING A CLUE

Mold is a kind of **fungus**. It grows from spores (tiny cells with hard coverings) that are so small that they float on air. Even though you can't see them, spores are already on the bread before you put it in the bag. By putting water on the bread and placing it in a cupboard, you are creating the warm, damp environment that helps spores grow. Unlike green plants, fungi do not make their own food from sunlight, air, and water. Fungi live by absorbing food from other sources. *Fungus* is from a Latin term meaning "food-robbing."

DISGUSTING DETAILS

Molds can give foods a strong, musky odor and make them taste very bad, but many people think that certain molds add flavor and texture to some foods. If you like Roquefort, a type of blue cheese, you are eating moldy cheese.

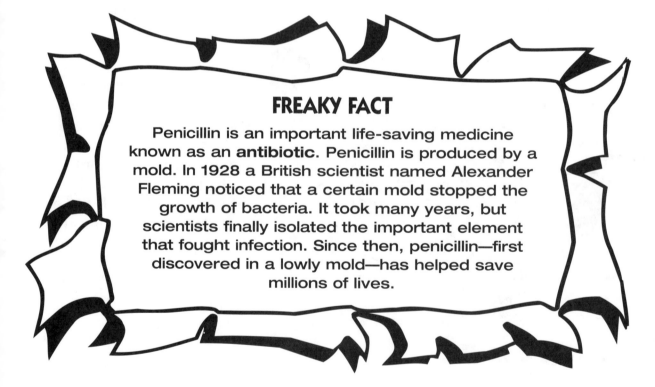

FREAKY FACT

Penicillin is an important life-saving medicine known as an **antibiotic**. Penicillin is produced by a mold. In 1928 a British scientist named Alexander Fleming noticed that a certain mold stopped the growth of bacteria. It took many years, but scientists finally isolated the important element that fought infection. Since then, penicillin—first discovered in a lowly mold—has helped save millions of lives.

ATTACK OF THE NIGHT CRAWLERS

Time Required:
Preparation – 2 to 3 hours;
2 to 3 months to complete

ADULT SUPERVISION NEEDED WHEN USING A HAMMER AND NAIL

Believe it or not, you can help the environment by raising a crop of squirmy houseguests that are first-class recyclers. They eat nothing but garbage and leave behind something very useful for the garden: compost. Made up of decayed plant matter, compost is one of the best fertilizers you can find.

STUFF YOU'LL NEED

- hammer and nail
- plastic bin with tight-fitting lid (approximately 1 foot x 2 feet x 6 inches)
- base material (grass clippings, finely shredded newspaper, dried leaves)
- water
- 1 cup sand or soil
- approximately 24 earthworms
- scraps of fruit, vegetables, coffee grounds, eggshells, tea bags (Do not use meat or dairy products, or anything with sugar or salt in it.)

LET'S DO IT!

1 With an adult's help, use the hammer and nail to poke 2 rows of holes in the sides of the container (at least 10 per side). Poke 10 or 12 holes in the bottom, too.

2 Gather enough base material to fill the container a little more than half full. Dampen the material slightly with water, then, with your hands, mix in a cup of sand or soil. Be sure the material is situated loosely in the container, not packed.

3 Place about two dozen earthworms, or night crawlers, into the container. (You can dig them up outside, or buy them at a store that sells live bait.) Set the container on blocks or bricks so that air can circulate around it. If you store it outside, find an area that is protected from wind, rain, and direct sunlight. It should also be in an area that won't be disturbed.

4 Add approximately 2 cups of food scraps by lifting chunks of the base material and slipping scraps within. Put on the lid. As the worms use up the food, you'll need to add more. Each time you do, put the food in a different area of the bin. Gently blend the scraps into the base material with your hands, being careful not to harm the tenants. Always wash your hands thoroughly after mixing. Repeat this step for 2 months. At that point you should find that much of the material in the bin has been converted into a dark, spongy "soil."

After 2 or 3 months your compost will be ready to use in a garden or in potted plants. Carefully remove the worms and release them in a safe place such as a garden or park, or put them aside while you set up a new batch of material for compost.

GETTING A CLUE

The earthworm is one of the original recyclers. It eats its own weight in decaying plant matter each day and passes out digested material, creating the world's best natural fertilizer. Another plus: Earthworms aerate the soil and create better drainage.

During the course of this experiment, if you feed your wiggly crew regularly and keep them in a place with good air circulation, the compost won't have anything but a rich, earthy smell. If it's stinky, you are probably giving the worms more scraps than they can handle. By the way, earthworms break down food scraps seven times faster than the material would break down without worms.

FREAKY FACT

Some people are squeamish about handling earthworms. That's easy to understand if you live in Australia. One earthworm species there can grow to 11 feet in length!

BUCKET OF BACTERIA

Time Required:
3 to 4 days to complete

ADULT SUPERVISION NEEDED WHEN USING STOVE

The world would be a much stinkier place if we didn't bathe or brush our teeth. Here's a way to see why.

LET'S DO IT!

1 Bring the water to a boil in saucepan, then, after it cools for a few minutes, pour carefully into the bowl.

2 Sprinkle the 4 packets of gelatin into the bowl and stir until the powder dissolves.

3 Pour the mixture into the jar, screw the lid on tightly, and set the jar on its side in a safe spot where it won't roll. Wait about 4 or 5 hours for the gelatin to set. (It is not necessary to put it in the refrigerator.)

4 Now you need to find some bacteria. You won't have to look far! Take off your shoes and socks and rub the cotton swab between your toes. Now brush the swab across the gelatin in four long, separate  strokes, then close the jar again and leave it in a warm, dark place for a few days.

 **5** When you check your experiment, you won't see the bacteria, but you will see lines in the gelatin. The gelatin is what the bacteria have been eating. Now open the jar and stand back. The contents of the jar will smell terrible!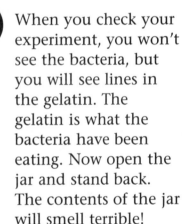

 6 When you are finished, dump the gelatin down the drain, followed by a rinse of hot water. Wash your hands and the inside of the jar thoroughly with soap and water.

 7 Repeat the steps again, but use a different source of bacteria. How about from your mouth or hands? Do you think you could collect bacteria from just-washed hands?

GETTING A CLUE

There are billions of tiny organisms living on and around us called *bacteria*. Some are helpful, others are harmful, and still others are somewhere in between. For example, some bacteria on humans are responsible for unpleasant odors in the mouth, underarms, and feet.

The gelatin you prepared makes a good food source for such bacteria. It is called a *growth medium*. It allows the bacteria to reproduce easily, and you get the concentrated effect when you open the jar to smell it.

DISGUSTING DETAILS

Bacteria are microscopic, single-celled **organisms.** Some are so tiny that it would take 50,000 of them to cover a square inch. They can also reproduce rapidly. Under ideal conditions, a single bacterium could possibly produce 3 billion new cells in a single day!

FREAKY FACT

Some very deadly diseases are caused by bacteria, including tetanus, leprosy, scarlet fever, and tuberculosis. On the other hand, life as we know it could not exist without certain bacteria, such as those that live in the human digestive system and help break down our food.

PUTRID, PUNGENT PUNCH

Time Required:
Approximately 1 hour to complete

If you can stand the stench, this foul-smelling fluid is a useful acid-testing indicator.

LET'S DO IT!

1 With an adult's help, cut the cabbage into small, bite-sized pieces and place in a pot. Pour in the distilled water and bring to a full boil. Turn off the heat and allow the fluid to cool. (The smell of this experiment will become apparent here!)

STUFF YOU'LL NEED

knife

half of a red cabbage

pot

1 quart distilled water

colander or sieve

measuring cup

3 small glass jars with lids

label

marker

measuring spoons

1 tablespoon lemon juice

¼ teaspoon baking soda

2 Pour the cooled liquid through a colander into a measuring cup. (You can use the cabbage in your compost heap [see page 56], or eat it with a little butter and salt. It is a good source of vitamins A and C, as well as calcium. Besides, it tastes great.)

The cabbage water will be an inky blue. Pour it into the first glass jar and label it *cabbage indicator.*

 Pour about ¼ cup of cabbage indicator into another glass jar, then add the tablespoon of lemon juice. The indicator will turn bright pink in the presence of acid.

 Pour about a ¼ cup of cabbage indicator into the last glass jar. Add ¼ teaspoon of baking soda. The indicator turns dark green in the presence of a base (a substance that reacts with acid to form a salt).

 Using the same process, test other foods to see if you can detect an acid or a base. Coffee, vinegar, and milk are good foods to test.

GETTING A CLUE

The hot water releases chemicals from the cabbage, some of which appear blue. When the chemicals come in contact with an acid or a **base**, a chemical reaction changes the color of the fluid from blue to pink or green. By carefully watching for any changes in the color of the indicator, you can detect even small amounts of acid or base in the substance you are testing.

GLOSSARY

acid: A chemical compound containing hydrogen.

air pressure: The force exerted by the air in the atmosphere. It is approximately 15 pounds per square inch at sea level at 32°F.

aldehydes: A class of highly reactive organic chemicals.

antibiotic: A drug used to treat infections caused by bacteria or fungi.

antioxidants: Substances that prevent or delay chemical interaction with oxygen. Vitamin C is a natural antioxidant.

atom: The smallest unit of matter that cannot be divided by a chemical reaction.

bacteria: Single-celled, microscopic organisms. One of these organisms is called a *bacterium*.

base: A substance that reacts with an acid to form a salt.

capillary action: The rise or fall of liquids in narrow tubes.

centripetal force: A force that acts on a moving body that causes that body to move in a curved path.

chemical energy: Energy released by a chemical reaction.

chemical reaction: The interaction of two or more substances, which results in chemical changes in the substances.

compost: Decayed plant material that provides nutrients in the soil for growing plants.

concentration: The amount per unit volume of one substance in a given space or other substance.

contracts: Becomes smaller.

denatured: What happens when a protein is changed, chemically or by heat, causing the protein to lose its solubility.

density: The mass (amount of matter) of a substance (per unit volume).

embryo: Inside a plant seed, the first stage of a new plant, before it has begun to grow.

energy: The ability to do work.

enzyme: A protein in the body that helps a chemical reaction take place.

evaporates: Changes from a liquid to a gas through heat or air movement.

fluids: Substances that flow, such as liquid or gas. Fluids take the shape of the container holding them.

frequency: In sound waves, the number of waves per second.

friction: The resistance to motion between two surfaces.

fungus: A plantlike organism that has no chlorophyll and cannot produce its own food. Mushrooms, mold, and yeast are examples of fungi.

heat: A form of energy transferred from one body or substance to another of lower temperature. Heat is released by physical and chemical processes.

immiscible liquids: Liquids that do not mix together, such as oil and water.

kinetic energy: Energy possessed by a body by virtue of its motion.

lubricant: A substance used to reduce friction between moving parts. Oil is a lubricant.

membrane: A thin layer of tissue that surrounds, separates, and protects living cells. Plant and animal organs may be separated or lined by a membrane.

metamorphosis: A change from one state to another in the life cycle of certain organisms.

microorganisms: Living things too small to be seen without a microscope.

molecule: The smallest unit of a compound that can exist by itself and still have the properties of that compound.

organisms: Living things.

osmosis: The passage of a solvent from a less concentrated into a more concentrated solution through a membrane that only partly permits the liquids to pass through it.

photosynthesis: The process by which green plants produce food from air, water, and sunlight.

pitch: The highness or lowness of a sound. The greater the frequency of the sound waves, the higher the pitch.

potential energy: The energy a body or object possesses by virtue of its position.

protein: An organic compound containing amino acids. Proteins are necessary for the body to function.

soluble: Able to be dissolved in a liquid.

solution: A liquid, gas, or solid in which one or more substances are dissolved.

sound: A vibration that travels through a gas, liquid, or solid, but not through empty space. The human ear detects the vibrations, and the human brain interprets the vibrations as sound.

spore: The reproductive cell of certain organisms such as fungi and some plants.

suspension: Particles of matter spread throughout but not dissolved in a liquid.

toxic: Poisonous.

vegetative propagation: The use of part of a plant, other than a seed, to produce a new plant.

viscosity: The resistance to movement within a fluid.

volume: The amount of space taken up by a substance or object.